THE LIBRARY OF
FUTURE ENERGY

SOLAR POWER
OF THE FUTURE
NEW WAYS OF TURNING
SUNLIGHT INTO ENERGY

SUSAN JONES

THE ROSEN PUBLISHING GROUP, INC.
NEW YORK

Published in 2003 by The Rosen Publishing Group, Inc.
29 East 21st Street, New York, NY 10010

First Edition

Library of Congress Cataloging-in-Publication Data

Jones, Susan, 1965–
Solar power of the future: new ways of turning sunlight into energy/
Susan Jones.
    p. cm.— (Library of future energy)
Summary: Discusses various kinds of solar energy, the history and
development of their use, economic aspects of solar energy, and future
possibilities.
Includes bibliographical references and index.
ISBN 0-8239-3663-5 (library binding)
1. Solar energy—Juvenile literature. 2. Photovoltaic power generation—
Juvenile literature. [1. Solar energy. 2. Power resources.]
I. Title. II. Series.
TJ810.3 .W45 2002
333.792'3—dc21

                                                    2002000224

Manufactured in the United States of America

CONTENTS

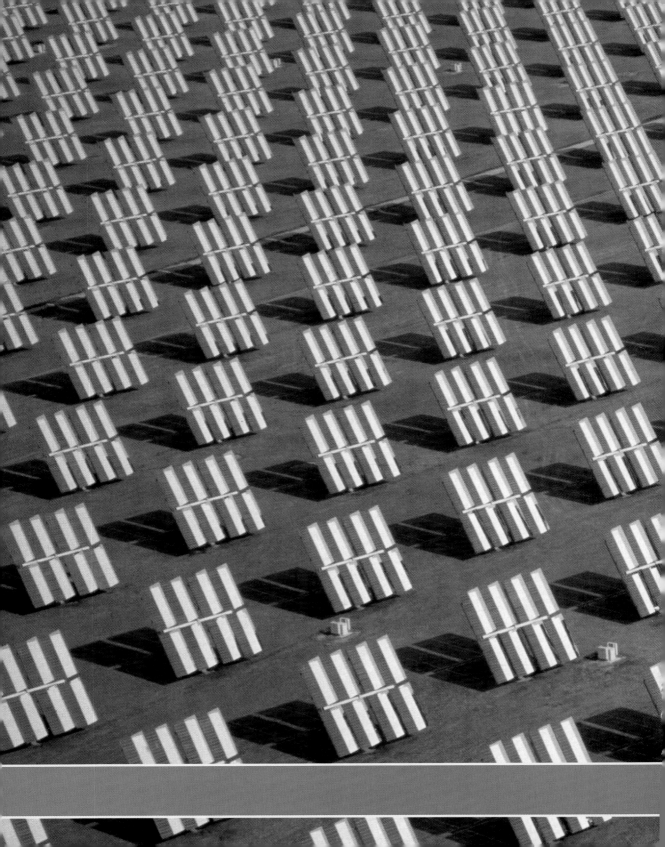

INTRODUCTION

Have you ever wondered what heats or cools your home, or what powers your lights, television, and computer? It's probably a fossil fuel like oil, gas, or coal. Most of us live comfortably because of fossil fuels, which, until recently we have taken for granted.

Did you know that it takes millions of years for fossil fuels to form underground and it takes billions of dollars to remove fossil fuels from the ground and convert them to power? And the fossil fuels we use to heat water for a hot shower or a load of laundry leave behind dangerous pollutants. Pollutants poison our environment, kill off plants and animals, and make many of us sick. Despite all the dangers, fossil fuels are still the most popular source of power in the United States.

This house in Maryland generates its own electricity and hot water using free energy from the Sun, which is harvested through a special roof equipped with solar energy panels.

Rather than depend on fossil fuels to produce heat, some people now use the Sun to do the same thing. Have you ever noticed panels that look like windows on the roofs of some buildings? They are solar energy panels that capture sunlight and turn it into energy.

An amazing thing about solar energy is that it is renewable. It re-forms quickly and leaves behind no pollutants. Many people believe that to protect our environment from destruction we must learn to use solar energy and other renewables, like wind energy, to meet our future heat and electricity needs. People in science, industry, and government are experimenting with ways to use solar energy more efficiently.

FOSSIL FUELS

At the turn of the twentieth century, scientists began to warn that the smoky soot released by burning coal was toxic. They said it would make us sick, injure the environment, and change our weather systems. Not many people paid attention.

In the last thirty years, more and more scientists have recognized that the early warnings were on target. Fossil fuels were sending poisons into the air that were ruining Earth's important ecosystems. Our climate has been getting warmer as a result of greenhouse gases released by some fossil fuels. Unusual floods and strange weather patterns are occurring around the world. Who is being hurt by this? People are, as well as plants and animals!

We use huge amounts of fossil fuels every day to provide us with electricity and heat. According to research done in 1992, if California were a separate nation, it would be the world's fourth largest consumer of energy. Nine years later, in June 2001, the state of California could not generate enough energy to meet its demand. Many homes and businesses lost power. Without electricity, people could not use the Internet or watch television. They could not turn on lights or take hot showers. People went grocery shopping with flashlights!

SOLAR ENERGY

But there is good news. Scientists have been researching alternative ways to create energy for heat and electricity using renewable sources. Like fossil fuels, renewable energy forms naturally. But renewable energy regenerates, or forms again, very quickly. This is a big plus. Solar power, which uses sunlight, is a type of renewable energy that many scientists think will be an important energy source in the future.

Seventy percent of Americans now see global warming and climate change as a danger to the environment. Three-quarters of Americans want to decrease the United States's oil purchases from foreign countries. Americans are even willing to spend more money on renewable energy to lessen dependency on foreign countries for oil.

Concern for the well-being of the earth reaches every age group. In 1990, at the Bergen Conference in Norway, children voiced these concerns and ideas to politicians and world leaders:

- We want to inherit a clean earth.

- Why shouldn't we be allowed to live as you did when you were little?

- We want to play in fresh forests, fish in any water, and drink clean water from the river.

- Life is more important than money.

- We must find energy that does not pollute.

To improve the quality of our air and protect our environment, the U.S. government will have to reshape its policies. Right now, most solar energy equipment made in the United States is exported, or sent overseas, to nations like Germany, the Netherlands, and Sweden. These countries are investing heavily in renewable energy technologies. While the United States explores ways to use the light of the Sun for its energy needs, it remains committed to using oil, gas, and coal, as well as nuclear power.

But there are positive changes in the United States. During his presidency, Bill Clinton announced a project called the Million Solar Roofs Initiative. The goal of this project is to equip one million homes and businesses in the United States with solar roofs by 2010. Projects like this will encourage change.

In this book, you will learn about many projects in the United States that are demonstrating effective ways to use solar energy in the future. We will all be better off when we depend less on fossil fuels and more on solar energy and other renewables.

The Sun, a powerful source of energy, is responsible for nearly all the energy on Earth except for the moon tides, radioactive material, and Earth's internal heat.

A fiery star, the Sun has been burning for over four billion years. The needs of people account for only 0.1 percent of the Sun's total power. In fact, enough sunlight reaches Earth each minute to meet the world energy demand for a year.

RENEWABLE ENERGIES

Renewable energies are power sources that humans cannot deplete or use up. As we use them, they develop again. The Sun, one of our most powerful sources of renewable energy, burns constantly. It replaces itself rapidly. Wind, water, and hydrogen power are other types of renewable energy.

The other good news about renewable energy is that it's safe for the environment. Renewable energy doesn't release poisonous particles or dangerous fumes into the air. Most leave behind little or no hazardous materials.

Even so, in 1994, renewable energy accounted for only 11 percent of the United States's electricity. Of the electricity used worldwide, 18 percent was produced from renewable energy. If only 11 percent of U.S. electricity comes from renewable energies, can you guess where the rest of it comes from? That's right. It comes from fossil fuels.

Fossil Fuels

For a long time, we have relied on fossil fuels to produce energy. Like renewable energies, fossil fuels are formed naturally. But unlike renewable energies, it takes millions of years for fossil fuels to form in the earth. Not only is it expensive to remove fossil fuels from the earth, the process causes damage to land, water, and wildlife habitats. This damage is very costly to repair and, in some cases, the damage cannot be repaired. But the fossil fuel industry is so large in the United States that without it many people would lose their jobs.

THE BASICS OF SOLAR ENERGY

We use several methods to harness the power of the Sun. First, we can generate electricity directly from sunlight. We call this photovoltaic (PV) technology. We can also use the Sun's heat to raise the temperature of a fluid. When the fluid becomes boiling hot, it powers a machine that makes electricity. We call this solar thermal technology. We can also design and construct buildings to absorb and slowly release heat from the Sun.

PHOTOVOLTAIC OR SOLAR ELECTRIC SYSTEMS

Photovoltaic or solar electric systems are fueled by sunlight. Solar cells are called photovoltaic cells. They are made from thin slices of silicon, a plentiful natural resource that is also a major component in computers.

Close-up of single-crystal photovoltaic cells. Photovoltaic systems were used exclusively in space exploration before researchers applied the technology to commercial and residential use.

Photovoltaic cells are the building blocks of solar electric systems. They change some of the energy directly into electricity.

There are several advantages to using solar electric power. Since solar electric power converts sunlight directly to electricity, it does not need a bulky generator system. Without a generator to cool, photovoltaic cells have no poisonous liquids, no caustic chemicals, and no moving parts. As a result, the environmental impact of a photovoltaic system is small compared to a coal-burning power plant that sends deadly mercury into the air.

PHOTOVOLTAIC CELLS

The photovoltaic cell, the basic unit of a photovoltaic system, looks like a flat sandwich. It can be tiny enough to power a watch or calculator. Or a photovoltaic cell can be as large as four inches across. Since one cell produces only one or two watts of energy, for big jobs, cells are joined together to make modules. PV modules, which have no moving parts, can last from twenty to thirty years.

PV modules are wired together to form a panel. A collection of panels is called an array. Arrays are grouped into array fields. A complete PV system includes array fields, charge controllers, storage batteries, and tracking equipment.

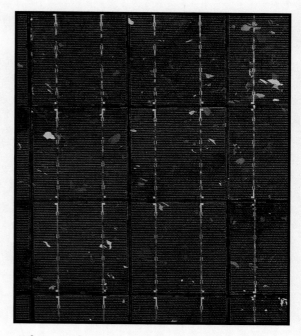

Solar panels convert radiant energy from the Sun into electrical energy.

Sunlight is made up of photons, or particles of solar energy. When photons strike a solar cell, several things can happen. Some photons are reflected by the cell, some photons pass through the solar cell, and some photons are absorbed by the cell. Only the absorbed photons can provide the energy to generate electricity.

When a cell absorbs enough sunlight, the photons release electrons, atomic particles with negative charges. The electrons move to the surfaces of the cell. The negative electrons on the cell create an imbalance between the front and back surfaces. It's similar to the positive and negative ends of a battery. When the two surfaces of the cell are connected by an external piece, electricity flows. This phenomenon, called the photovoltaic effect, is what makes solar electric power.

Solar electric power depends on sunlight, and lack of sunlight can be its biggest drawback. When the weather is cloudy, cells can't catch enough sunlight to produce sufficient amounts of electricity to make things work. Scientists are creating ways to solve this problem.

SOLAR THERMAL SYSTEMS

Solar thermal equipment captures heat and transfers it to a fluid. Most solar thermal systems are used to heat buildings and water. Solar thermal equipment is also used to generate electricity, dry crops, and destroy dangerous waste.

Right now there are about 1.2 million solar thermal systems operating in the United States. Over 80 percent of these systems have been installed in people's homes, usually to heat pools.

SOLAR THERMAL COLLECTORS

If you want to catch a butterfly, you use a net. To catch sunlight, you use a solar thermal collector such as a roof panel to absorb the rays of the Sun.

Solar thermal technology is extremely popular, but there are still challenges. For example, how large do the panels have to be to capture sunlight? How do the roof panels capture light on a rainy day or when the sky is cloudy? Scientists are developing a variety of solutions.

There are three types of solar thermal collectors: low-temperature, medium-temperature, and high-temperature.

MEASURING ENERGY

Electrical energy is measured in amperes, volts, and watts. An average household uses around 700 kilowatts per month or about 8,500 kilowatts per year.

- Amperes measure electrical current.

- Volts measure electrical pressure.

- Watts measure electrical power (amperes × volts).

  - A kilowatt (kW) is 1,000 watts.

  - A megawatt (MW) is one million watts or 1,000 kWs.

  - A gigawatt (GW) is one billion watts or 1,000 MWs.

Low-temperature solar thermal collectors provide low-grade heat—less than 110 degrees Fahrenheit. People use this system mostly to heat their swimming pools.

Medium-temperature solar thermal collectors produce medium- to high-grade temperatures, usually around 140 to 180 degrees Fahrenheit. Medium-temperature collectors are used mostly to heat hot water for homes.

High-temperature solar thermal collectors are the most powerful. They use huge, bowl-shaped structures called parabolic dishes. High-temperature collectors are used mostly by power companies to generate large amounts of electricity.

Some people back up their concern about global warming by using solar collectors, like the one pictured here, to heat the water in their homes.

SOLAR BUILDING OR PASSIVE SOLAR POWER

Solar building, or passive solar power, is the simplest and oldest type of solar energy. Passive solar power is brought into play when a building is constructed to gather the most amount of sunlight. This can be as simple as placing large windows along the south side of a building. Windows that face the south receive the most light and warmth from the Sun.

Using passive solar energy may also mean using materials for a building's floors and walls that absorb and store the natural sunlight. As they are hit by the rays of the Sun, the walls and floors heat up. When it gets cold at night, the walls and floors slowly release the heat. This is called direct gain.

TYPES OF SOLAR BUILDING DESIGN

Other types of solar building are sunspaces, Trombe walls, and day-lighting. A sunspace, which looks like a greenhouse, is built into the south side of a building. As sunlight comes through the glass, it warms the sunspace. A special ventilation system circulates the heat through the building to where it is needed.

In contrast, the Trombe wall is a thick, south-facing wall, which is painted black. Black objects absorb the most amount of sunlight, and the Trombe wall is made of materials that absorb enormous amounts of heat.

With daylighting, a building is constructed to allow for the most amount of natural light. Very often, a row of windows, called clerestory windows, is installed near the peak of the roof to bring in more sunlight.

Buildings that incorporate passive solar designs can cost as much as 50 percent less to heat than traditional buildings that are not energy efficient. Passive solar designs can also include natural ventilation for cooling.

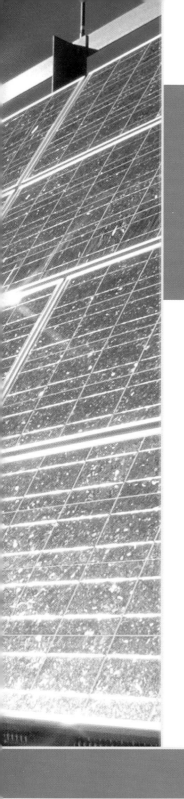

2 HISTORY AND DEVELOPMENT

People have been using renewable energies much longer than they have been using fossil fuels. Solar building dates all the way back to the year AD 100! Solar thermal methods, while not nearly as old as solar building, are more popular today.

THE EVOLUTION OF SOLAR THERMAL ENERGY

In 1767, a Swiss scientist named Horace de Saussure built the world's first solar collector. In the United States, the father of solar energy was Clarence Kemp, an inventor based in Baltimore, Maryland. In 1891, Kemp patented the first commercial solar water heater, making him the only person allowed to build and sell this solar design. Four years later, in 1895, two Pasadena, California, business tycoons bought the rights to Kemp's solar energy invention.

By 1897, 30 percent of Pasadena homes were equipped with solar water-heating systems. The people who had purchased the heaters had many of the same incentives then that people have now. They were tired of paying high prices for gas and coal.

Expanding Demand for Solar Thermal Energy

In 1908, William J. Bailey, of the Carnegie Steel Company, invented a solar thermal collector with an insulated box and copper coils. Bailey's design is remarkably close to the design used today. No one has found a better method to collect energy from the Sun.

Moderately successful with his invention, by 1918 Bailey had sold about 4,000 units. After Bailey sold the patent rights to a businessman in Florida, nearly 60,000 units were in use by the end of 1941. However, production fell when copper was rationed during World War II since copper coils were one of the two main components of the collector.

Ups and Downs

In the 1960s, solar energy in the United States suffered another blow. Solar water heaters were being manufactured when President Richard Nixon passed a bill that allowed the United States to import

THE HOT BOX

In 1767, Swiss scientist Horace de Saussure built the world's first solar collector. Called a hot box, it was a nest of many small boxes. De Saussure placed the hot box in the sun, making sure the sunlight hit it directly. After a while, he measured the temperature and found it had increased as the boxes got smaller. The heat had reached 189.5°F.

In the 1830s, astronomer Sir John Heschel used de Saussure's solar hot boxes to cook food during an expedition to southern Africa. Solar thermal energy became important in parts of Africa for purifying water, through heating, for drinking and cooking.

50 percent of its oil. When oil became more plentiful and cheaper, people became less interested in solar and other renewable energies.

But the interest in solar energy picked up again as a result of the oil embargo of 1973. The members of OPEC (the Organization of Petroleum Exporting Countries) raised the price of oil and cut production. The OPEC countries include Iran, Iraq, Kuwait, Saudi Arabia, Venezuela, Qatar, Indonesia, Libya, United Arab Emirates, Algeria, and Nigeria. Gasoline and oil became extremely expensive in the United States, and there wasn't enough of it to go around. Americans began to realize that they needed to look for new ways to heat their homes, schools, and businesses. When President Jimmy Carter had solar energy panels installed on the White House during this time, people began to get interested in using solar energy once again.

In 1974, a California company called FAFCO and a Colorado company called Solaron became the first companies in many years to manufacture solar heating equipment on a large scale. By 1985, other solar energy manufacturers had emerged. Soon, sales of solar energy equipment had reached $800 million. It was a big leap forward for solar thermal technologies!

At one time, many solar energy manufacturers received tax credits for providing alternative energy systems. Today, those tax credits have almost disappeared. But there are still a handful of solar thermal manufacturers in the United States that produce equipment for more than 1.2 million buildings in the United States that have solar water-heating systems, and 250,000 swimming pools that are equipped with solar heating systems.

SOLAR ELECTRIC ENERGY: LEARNING TAKES TIME

In 1839, Edmund Becqueral, a French physicist, first observed the photovoltaic effect, a discovery that allows solar electric energy to exist today. In 1883, American inventor Charles Fritts created the first solar cells from selenium. The early cells, which could convert light into electricity, were very weak. Because they captured very few electrons, they did not convert very much sunlight into energy.

It wasn't until 1954 that Bell Telephone Laboratories produced a silicon PV cell that was 4 percent efficient. Later, scientists at Bell Labs learned how to make a PV cell that was 11 percent efficient, a more useful amount.

In 1958, the Vanguard space program, directed by the National Aeronautics and Space Administration (NASA), used a PV cell to power its satellite's radio. The space program has been involved in the development of PV technology ever since. Its funding and research has helped scientists gather important information about PV systems for use on Earth.

President Jimmy Carter speaking at the dedication ceremony for a new solar heating system in the White House on June 20, 1979. By installing the system at the White House, Carter hoped to raise awareness of the need for new energy sources.

THE OIL EMBARGO CREATES SHORT-LIVED INTEREST IN PVS

In 1973 and 1974, during the oil embargo, the U.S. Department of Energy funded the Federal Photovoltaic Utilization Program, which enabled the installation and testing of over 3,100 PV systems. Most of these systems are still operating today.

But interest in solar power was dulled once oil prices dropped again. Neither Ronald Reagan nor George H. W. Bush paid much attention to renewable energy during their presidencies.

But the status of PV systems and solar energy improved greatly during the Clinton administration, as President Clinton began the forward-thinking Million Solar Roofs Initiative.

THE MILLION SOLAR ROOFS INITIATIVE

On June 26, 1997, during a conference for the United Nations Session on Environment and Development, President Clinton announced one of the biggest steps toward changing our current energy path. It was the largest energy initiative to be supported by the federal government in thirty years. The goal of the plan was to install photovoltaic energy systems in one million U.S. buildings by 2010.

In October 2000, Department of Energy officials announced that more than 100,000 solar energy systems had been installed since the beginning of the Million Solar Roofs Initiative.

SOLAR BUILDING

In AD 100, a writer named Pliny the Younger completed the first solar building. Not knowing he was inventing a revolutionary form of solar energy, Pliny added a room to his summer home in Italy that featured windows on all the walls. The room, which became hotter than the other rooms in the house, stayed warm during the

President Clinton, shown here speaking at an Earth Day event in 2000 at Sequoia National Park in California, launched a number of initiatives during his administration that bolstered the use of solar energy systems throughout the United States.

cold evenings. The warmth provided by sunlight saved him a lot of money on wood for heat.

In the fourth century AD, all bathhouses in Rome had south-facing windows to let in the warmth of the Sun. By the sixth century, sun-rooms had become so popular in homes that laws, called sun rights, were passed. The laws made sure that everyone had equal access to the Sun. Laws like these are still intact today, especially in the big cities of Japan, where tall buildings often cast shadows over smaller buildings.

In more modern times, architects have been responsible for the creation of solar building and passive solar construction. The design and position of a building play the biggest role in capturing sunlight.

Building a home with a row of windows for a wall is a great way to store the heat from sunlight.

THE 1800S AND BEYOND

During the 1800s, conservatories filled with exotic plants were very popular. Visitors flocked to these glass-walled buildings, which were flooded with sunlight and natural heat.

By 1947, passive solar buildings were in such demand that the book *Your Solar House* was published. The book profiled forty-nine of the United States's best solar architects, including Buckminster Fuller.

Passive solar heat was used exclusively in homes until 1951, when Frank Bridgers designed the world's first commercial solar

BUCKMINSTER FULLER

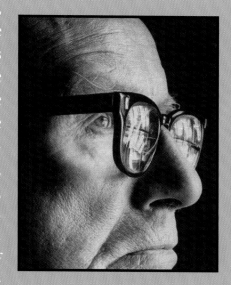

In 1927, at the age of thirty-two, Buckminster Fuller stood at Lake Michigan and thought about taking his life. His young son had just died, and he was so full of grief he thought suicide was the only answer. It was then that he realized how much a part of the universe he was. From that day forward, Buckminster Fuller lived out his beliefs, earning forty-seven honorary doctorates for his work in science, engineering, the arts, and the humanities. One of his greatest creations was the geodesic dome. It is the lightest, strongest, and most cost-effective structure ever designed. Buckminster Fuller believed passionately in renewable resources. He used the concept of solar building in all of his designs.

building. Called the Bridgers-Paxton Building, it still operates today and uses solar water heating and passive solar design.

In the mid 1990s there was another surge of interest in solar building design, particularly in the world of architecture. Currently, the American Institute of Architects (AIA) is working with other associations to establish programs that support solar building design.

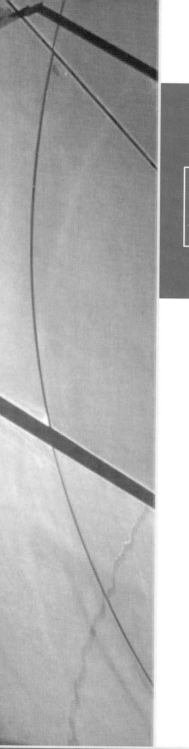

3 THE BUSINESS OF SOLAR ENERGY

Over the years, the United States's interest in solar energy has seesawed back and forth. When oil was plentiful and cheap, no one cared much about renewable energy. During the oil embargo of the 1970s, interest in renewable energy surged as the price of oil soared. No matter how great an invention or idea is, it is always a challenge to make it a commercial success.

The business of solar energy can be frustrating. Although making money is the goal of business, it has not always been in the best interest of the environment or of people.

Two billion people around the world live without electric power because they are too poor to pay for it. They have no heat, no

Under its Renewable Energy Rural Electrification Project, the government of Brazil provides electricity to many remote rural communities by installing solar panels in homes.

lights, and no clean drinking water! They suffer from many more diseases than people in the developed world. Many children do not live to be adults, and most adults do not live very long because they can't kill the dangerous germs in the water by boiling it.

- In Africa, 50 percent of the people do not have power.

- In Brazil, 20 percent of the population does not have electricity.

- In Central America, 55 percent of people lack power.

- In China, 50 percent of the residents have no electricity.

- In India, 72,000 villages do not have power.

- In Indonesia, 36,000 villages are without electricity.

FOSSIL FUEL OWNERS

You might wonder why so many people throughout the world lack electric power. There are no easy answers. The corporations that own the oil fields and coal mines, both in North America and abroad, and the organizations that run the plants that produce and distribute the power, are wealthy and influential. They want to continue to profit by providing these products and services. Very poor people do not have the money to pay their fees.

The federal Energy Information Administration (EIA) expects America to be importing 60 percent of its oil at a cost of $100 billion per year by 2006. This will increase the price of oil for everyone in the United States.

SOLAR ENERGY OWNERS

The business of solar energy is very different from the business of fossil fuel for one important reason. Corporations can sell oil because they own large oil fields, but no one can own the Sun. So what can be owned in the business of solar energy?

Solar panel manufacturers tend to have the biggest opportunity for profit when and if solar energy takes off. Without solar panels, there is no way to capture sunlight and convert it into heat. Companies that produce solar cells may also become very successful. As interest in solar energy increases, countries are becoming more competitive in the manufacturing of panels and cells. In 1996, the United States was the world's largest manufacturer of photovoltaic equipment. But in 1999, Japan surpassed the United States.

REGULATION AND DISTRIBUTION OF SOLAR ENERGY

Each time the price of oil goes up, interest in solar energy goes up, too. As there are more threats of energy crises, and as there are more energy blackouts like those in California, the government will once again look at developing solar technologies. At this writing, there is not enough enthusiasm from the U.S. government to spur support for solar energy. Instead, the manufacturers of solar technology are spreading the word about the advantages of renewable energy. They want the public to know how solar energy can lower users' expenses without harming the environment.

EARTH DAY

On April 22, 1970, Senator Gaylord Nelson, of Wisconsin, and Denis Hayes, a Harvard law student, founded Earth Day, believing that it was very important to set aside a day to discuss the state of the environment. Earth Day, first and foremost, is a day when people

The Frederick, Maryland, headquarters of Solarex in 1997. Then the largest U.S.-based maker of solar energy products, the company has since merged with BP Solar to form one of the world's leading solar energy companies.

can learn about the value of recycling and how to use renewable energy. With slogans like "Earth Day Every Day," Earth Day promotes the use of alternate sources of energy. At parks and meeting areas around the world, more and more people learn each year how to take better care of the earth.

While the main function of Earth Day is to educate and inspire, it is also a business opportunity for people who sell environmentally friendly products. For consumers, it is a chance to purchase materials that will enable them to live without harming the air, the earth, or their bank accounts.

Participants carry a sign proclaiming the first Earth Day on April 22, 1970. Today the Earth Day Network, a worldwide alliance, coordinates Earth Day events in the United States and around the world. In 2000, the Earth Day theme in the United States was New Energy for a New Era.

FUTURE CONSUMPTION OF SOLAR ENERGY

Solar energy and other renewable energy technologies will be a very large part of our future. The Energy Information Administration predicts that the world's consumption of energy will increase from 96 quadrillion Btus in 1999 to 126 quadrillion Btus in 2020. (Btu stands for British thermal unit. It is the measurement used to determine how much heat is needed to raise the temperature of one pound of water by one degree Fahrenheit.)

That is a significant increase and a potentially dangerous increase. Most scientists think that, on top of depleting our fossil fuels, we will also create more environmental problems if we continue to depend on them.

The next chapter discusses the controversies that surround the United States's and the world's next moves toward a new energy path. Some scientists argue that using renewable energies on a large scale will be too expensive, while others completely disagree. Still other scientists and economists say that fossil fuels are not harming our environment. These issues will need to be resolved before new energy paths can be established.

4 MAKING A NEW ENERGY PATH

The road to developing new energy technologies is paved with controversy. Where do we go next? How do we get there? Some economists and fossil fuel lobbyists suggest that installing equipment that will hold carbon emissions steady could cost hundreds of billions of dollars, much more than business, industry, and the public can afford. At the same time, some scientists warn that the threats from fossil fuels to people, animals, and Earth are too dangerous to ignore. Environmentalists and other concerned citizens want us to switch to renewable energy technologies to protect life on Earth as we know it. They tell us that the cost to consumers and to the government will eventually balance out.

OUR PLANET'S TEMPERATURE

Global warming is the increase of the average temperature of Earth. Scientists think that global warming is caused by the buildup of carbon dioxide (CO_2) in the air that has been released by burning fossil fuels such as oil and coal. The carbon dioxide has trapped heat from the Sun that normally escapes to the upper atmosphere. According to *Scientific American* magazine, the higher temperatures on Earth are causing dangerous changes in climate and weather.

As a result of warmer temperatures, glaciers are beginning to melt, flooding many low-lying areas of the world. If this continues, areas once suitable for farming will be covered with water, or they will become too warm to farm. Buildings will be covered with water, too. Global warming is also beginning to cause more severe storms, droughts, and other weather conditions. And the warmer climate will eventually spread many serious illnesses.

POLLUTION AND YOUR HEALTH

The major air pollutants regulated by the Environmental Protection Agency (EPA) are sulfur dioxide, nitrogen oxides, ground-level ozone, lead, and carbon monoxide. The pollutants are sent into the air by the burning of fossil fuels. Power plants, cars, and trucks produce the largest amounts of carbon monoxide (CO) and nitrogen oxides (NO_2 and NO). These gases lead to ground-level ozone, which poisons the lower atmosphere in which we live.

Global warming is believed to be the cause of melting polar ice caps. Environmentalists propose that using more sources of renewable energy such as solar power, instead of fossil fuel, where possible, can slow the threat of global warming.

Ground-level ozone, also known as smog, can irritate your lungs and cause bronchitis. It can increase your chances of developing respiratory illnesses, such as asthma. Coal-burning plants emit two-thirds of the sulfur dioxide (SO_2) in the United States. Sulfur dioxide is the primary cause of acid rain, which damages or destroys crops, lakes, and rivers.

Many people in the United States have been made sick by pollution. In 1995, 33 million Americans lived in cities that failed to meet federal smog regulations. As of 1997, 50,000 people in the United States died from pollution-related illnesses.

Pollution and the Environment

Acid rain is a broad term that describes the ways acids fall out of the atmosphere, polluting large areas of the United States and Canada. Rain, fog, and snow all contain acids that damage crops, trees, and soil. Acid rain pollutes lakes, streams, and marshes so that fish and other wildlife can no longer survive.

Acid rain also eats away at irreplaceable historic buildings and statues. The sulfates and nitrates associated with acid rain that form in the air from fossil fuel emissions have cut down on how far we can see when we are outside.

Reducing Emissions with Solar Power

In contrast to fossil fuels that create so much air pollution, solar power is much cleaner to use because it releases little or no polluting chemicals or gases into the air.

Photovoltaic power and solar building produce zero emissions. For example, low-e windows enable people to warm their homes or offices without fossil fuel. Covered with special film, low-e windows allow warmth from the Sun to pass into a building. The windows also prevent most of the heat from escaping. Industry experts estimate that by 2015, enough people will be

Smog is a chemical reaction between volatile organic compounds and nitrogen oxides in the presence of sunlight and warm temperatures.

using this kind of technology for heat to cut down dramatically on the amount of fossil fuel that is needed.

Emissions from burning gasoline when we drive our cars cause tremendous amounts of air pollution. But there are people who are determined to harness the power of the Sun to fuel automobiles. Each year since 1997, the grueling 2300km (1,426 miles) SunRace for solar-powered cars takes place in Australia. Entrants from all over the world, including the United States and Japan, come to race. Although solar-powered vehicles are a long way from being available to most of us, race sponsors compare their event to the great air races between Paris and London in the early twentieth century that led to the development of commercial airplanes.

MAKING CHANGES

Renewable energies can cut down on pollution and keep us healthier, as well as reduce the amount of harm we do to the environment.

Renewable technologies create less pollution because they do not release dangerous gases. By using more renewable energies, the United States could reduce the poisonous emissions in the air, which would reduce pollution-related health problems.

The information in this book is meant to explain how our current energy path may be damaging North America and our world. Understanding this means you can work to change it.

There are many things you can do to help promote the move from fossil fuels to renewable energies. At home, you can discuss having low-e windows installed to cut down on your family's need for fossil fuels. At school, you can talk to your friends and teachers about putting energy panels on the roof of your school building. Or, next April 22, you, your friends, and your family can attend the Earth Day festivities in your town or in a place nearby. Earth Day is fun and educational, too! Together you will be able to help spread the word about cleaning up our environment and keeping it clean by using renewable technologies. You can help lead your community, and even perhaps our world, onto a newer, safer energy path!

NATIONAL SECURITY

By 2006, the United States may buy half of its oil from countries in the Persian Gulf, which means that these countries will make $250 billion from oil sales to the United States. Some U.S. scientists and government officials believe we could better use this money to develop a stronger military and military weapons. People do not want to buy oil from countries that support international terrorism.

Environmental workers clean a beach after the *Exxon Valdez* oil spill in Alaska.

ENVIRONMENTAL HAZARDS

Oils spills and industrial runoff, excess chemicals that are dumped into lakes and streams, are a serious threat to wildlife. In 1989, the tanker *Exxon Valdez*, carrying oil in Alaska, spilled its contents into the ocean, damaging one of the cleanest and purest environments in the world. The cost to clean up the spill, which took about ten years, was $1 billion. The damage to wildlife was devastating.

5 INTO THE FUTURE

The future of renewable technologies is full of adventure. Each idea that scientists explore and develop is more amazing than the last. As new technologies are being unveiled, people seem more ready to receive and use them.

SOLAR CONCENTRATORS

Solar concentration, an innovative approach to gathering sunlight, is used in solar thermal technology to generate heat to power turbines. Solar cells operate more efficiently when the Sun shines directly on them in a wide beam. With solar concentration technology, mirrors or lenses focus light onto specially designed cells. Heat sinks cool the cells.

Older versions of solar cells have solid absorbing layers that require clear skies and direct sunlight. They will not operate under cloudy conditions. But, new solar concentrators follow the Sun's path through the sky during the day, so they have a better chance of finding direct sunlight.

ELECTROCHEMICAL PV CELLS

Electrochemical solar cells have active absorbers that are in a liquid state. They also have a new layer called a dye sensitizer. With a glass or foil backing sheet, this layer is sandwiched between the front contact layer of the cell and the rear carbon contact layer. The new layer absorbs light and creates electron movement in the semiconductor layer. Some scientists and researchers believe that these cells will cost less to manufacture in the future because they are simple to construct from inexpensive materials.

BETTER PV CELLS

In June 2001, the *Denver Post* ran a story about a new breakthrough in renewable energy technology. It is a PV cell that is more efficient than its predecessor. Better efficiency means that the cell absorbs and creates more power. Xuanzhi Wu, a senior

FACTS ABOUT PV SYSTEMS

- The Department of Energy and the solar energy industry have reduced PV system costs by more than 300 percent since 1982.

- The number of U.S. companies producing PV panels has doubled since the late 1970s.

- The most widespread application of PV systems is in consumer products, which use less than one watt to operate.

- More than one billion handheld calculators, several million watches, and a couple million portable lights and battery chargers use PV cells for power.

- The worldwide PV industry has grown from sales of less than $2 million in 1975 to greater than $750 million in 1993.

- Reliability and lifetime of PV systems are steadily improving; PV manufacturers guarantee their products for up to twenty years.

- In 1994 more than 75 percent of the PV modules produced in the United States were exported, mostly to developing countries, where two billion people still live without electricity.

- PV systems are rapidly becoming the power supply of choice for small-power tools of 100 watts or less, such as calculators and TV remotes. More than 200,000 homes worldwide depend on PV systems to supply all of their electricity.

A scientist from the National Renewable Energy Laboratory shows off a solar panel to a group of farmers during a workshop on the cost and environmental benefits of wind and solar power.

scientist at the National Renewable Energy Laboratory in Golden, Colorado, helped create the newly designed solar electric cell. He and others at the lab say that it surpasses previous cells in the percentage of sunlight it converts to electricity: 16.4 percent as opposed to 15.8 percent. Though it may seem a small success, the new knowledge may lead to the development of a cell with an even greater efficiency.

COMING OF AGE

In January 2001, the Environment News Service reported that as part of the U.S. Solar in the Jungle Project, a one-kilowatt solar/battery system has been constructed at Camp Leakey, a research center in Borneo, Indonesia. Scientists at Camp Leakey are fighting to preserve orangutans from extinction. In the absence of a reliable local utility grid, PV technology has come to their rescue, providing a solar-powered system for lights and to charge batteries.

As of April 2001, the Dangling Rope Marina, a national recreation area in Utah, switched from diesel-fired generators to solar energy. Using a 115-kilowatt solar array with a propane back up, the marina reduced its annual emissions by 540 tons of carbon dioxide, 27,000 pounds of nitrous oxide, 2,000 pounds of sulfuric acid, and 5,183 pounds of carbon monoxide.

In November 2001, the City of Brockton, Massachusetts, received a $30,000 Brightfields grant from the U.S. Department of Energy to study the possibilities for environmental cleanup and redevelopment using solar energy. The Brightfields project is a partnership established by President George W. Bush to promote the use of clean power and to stimulate the economy.

Workers from the Los Angeles Department of Water and Power set up solar panels on a building in downtown Los Angeles.

The Dawn of a New Energy Path

More and more scientists are concluding that the time is right for the United States to begin a new energy path. Yet many scientists have been voicing the same opinion since the 1970s and earlier. Overall, what is going to change the energy path is a cohesive voice from consumers, scientists, energy providers, manufacturers, and the government. Right now, many groups are beginning to support a change. Though change may come slowly, with more and more people initiating small changes, there is hope that renewable energy technologies will play a significant role in our future.

GLOSSARY

array A collection of solar panels, made up of photovoltaic modules.

by-product Something produced in an industrial or biological process in addition to the principal product.

carbon dioxide (CO_2) A heavy, colorless gas that is breathed in and out by humans and animals and is absorbed by plants. It is also given off in the burning of fossil fuels.

consumers People who buy and use material goods.

direct gain When the walls and floors of a solar building heat up during the day and slowly release heat at night when it is needed.

electrons Particles that circle the nucleus of an atom and carry a negative charge.

embargo An order prohibiting or restricting commerce.

emissions Substances discharged into the air.

energy   Useable power.

fossil fuel   A fuel such as coal, oil, or natural gas that is formed in the earth from the remains of plants or animals.

lobbyist   A person who rallies for a chosen group or cause.

parabolic   Shaped like a bowl.

renewable energy   An energy resource replaced rapidly by natural processes. Solar and wind power are renewable energy sources.

solar building   A type of solar energy in which a building is designed to gather the most amount of sunlight; also called passive solar power.

solar electric   A type of solar energy that changes the Sun's energy directly into electricity using solar cells; also called photovoltaic, or PV energy.

solar energy   Power from the Sun.

solar panel   A basic building block of a solar energy system in which a rectangular piece of material acts like a battery charger.

solar thermal   A type of solar energy that captures heat and transfers it to a fluid.

FOR MORE INFORMATION

Alliance to Save Energy
1200 18th Street NW, Suite 900
Washington, DC 20036
(202) 857-0666
e-mail: info@ase.org
Web site: http://www.ase.org

American Solar Energy Society
2400 Central Avenue, Suite G-1
Boulder, CO 80301
(303) 443-3130
Web site: http://www.ases.org

Center for Energy Efficiency and Renewable Technologies (CEERT)
1100 Eleventh Street, Suite 311
Sacramento, CA 95814
(916) 442-7785
e-mail: info@ceert.org
Web site: http://www.cleanpower.org

Solar Electric Power Association
1800 M Street NW, Suite 300
Washington, DC 20036-5802
(202) 857-0898
e-mail: SolarElectricPower@ttcorp.com
Web site: http://www.solarelectricpower.org

Solar Energy Industries Association
1616 H Street NW, 8th Floor
Washington, DC 20006
(202) 628-7745
e-mail: info@seia.org
Web site: http://www.seia.org

U.S. Department of Energy
1000 Independence Avenue SW
Washington, DC 20585
(800) dial-DOE (342-5363)
Web site: http://www.energy.gov

In Canada

Canadian Renewable Fuels Association
31 Adelaide Street East
P.O. Box 398
Toronto, ON M5C 2J8
(416) 304-1324
e-mail: publicinfo@greenfuels.org
Web site: http://www.greenfuels.org

Canadian Solar Industries Association
2415 Holly Lane, Suite 250
Ottawa, ON K1V 7P2
(613) 736-9077
e-mail: info@canSIA.com
Web site: http://www.cansia.ca

Kortright Centre for Conservation
9550 Pine Valley Drive
Woodbridge, ON L4L 1A6
(905) 832-2289
e-mail: kcc@interlog.com
Web site: http://www.kortright.org

Office of Energy Efficiency
Natural Resources Canada
580 Booth Street, 18th Floor
Ottawa, ON K1A 0E4
(800) 387-2000
e-mail: general.oee@nrcan.gc.ca
Web site: http://oee.nrcan.gc.ca

Ontario Ministry of Environment and Energy
135 St. Clair Avenue W.
Toronto, ON M4V 1P5
(800) 565-4923
Web site: http://www.ene.gov.on.ca

Solar Energy Society of Canada Inc.
P.O. Box 33047
Cathedral P.O.
Regina, SK S4T 7X2
e-mail: info@solarenergysociety.ca
Web site: http://www.solarenergysociety.ca

WEB SITES

Due to the changing nature of Internet links, the Rosen Publishing Group, Inc., has developed an online list of Web sites related to the subject of this book. This site is updated regularly. Please use this link to access the list:

http://www.rosenlinks.com/lfe/sola/

FOR FURTHER READING

Arnold, Guy, and Peter Harper. *Facts on Water, Wind and Solar Power.* New York: Franklin Watts, 1990.

Brooke, Bob. *Solar Energy.* Broomhall, PA: Chelsea House Publishing, 1992.

Brown, Paul. *Energy and Resources.* New York: Franklin Watts, 1998.

Daley, Michael J. *At Home with the Sun: Solar Energy for Young Scientists.* Putney, VT: Professor Solar Press, 1995.

Graham, Ian S. *Solar Power.* New York: Raintree/Steck-Vaughn, 1999.

Kerrod, Robin. *The World's Energy Resources.* Albany, NY: Thomson Learning, 1993.

Spence, Margaret. *Solar Power.* New York: Franklin Watts, 1993.

Spurgeon, R. *Energy and Power.* Baltimore, MD: EDCP Publishers, 1998.

BIBLIOGRAPHY

Berman, Daniel M., and John T. O'Connor. *Who Owns the Sun: People, Politics, and the Struggle for a Solar Economy.* White River Junction, VT: Chelsea Green Publishing Company, 1997.

Chiras, Daniel D. *The Natural House.* Broomfield, CO: Real Goods Solar Living Books, 2000.

Davidson, Joel. *The New Solar Electric Home.* Boulder, CO: Aatec Publications, 1990.

Pielou, E.C. *The Energy of Nature.* Chicago: University of Chicago Press, 2001.

Schaffer, John, and Doug Pratt. *Solar Living Source Book: The Complete Guide to Renewable Energy Technologies and Sustainable Living.* Broomfield, CO: Real Goods Solar Living Books, 2001.

INDEX

CREDITS

ABOUT THE AUTHOR

Susan Jones is a production editor and freelance writer. She lives in Brooklyn, New York.

DESIGN AND LAYOUT

Thomas Forget

JV

DWIGHT D. EISENHOWER LIBRARY
537 TOTOWA ROAD
TOTOWA BOROUGH, NJ 07512